Electronics Terminology: A Concise Dictionary

By

Informatik Rezurch

Butterworth-Heinemann

Boston Oxford Melbourne Singapore Toronto
Munich New Delhi Tokyo

Electronics Terminology:
A Concise Dictionary

A collection of non-proprietary computer, semiconductor, electronic packaging & interconnection acronyms & glossary

Butterworth-Heinemann
A member of the Reed Elsevier group

Recognizing the importance of preserving what has been written, Butterworth-Heinemann prints its books on acid-free paper whenever possible.

Library of Congress Cataloguing-in-Publication Data

Electronics terminology : a concise dictionary / by Informatik Rezurch.
p. cm.
ISBN 0-7506-9751-2 (pbk.)
1. Electronics--Dictionaries. 2. Electronics--Acronyms
I. Informatik Rezurch.
TK7804. E333 1996
621.381'03--dc20 95-48217
CIP

British Library Cataloguing-in-Publication Data
A catalogue record for this book is available from the British Library.

The publisher offers discounts on bulk orders of this book.
For information, please write:

Manager of Special Sales
Butterworth-Heinemann
313 WAshington Street
Newton, MA 02158-1626

10 9 8 7 6 5 4 3 2 1

Printed in the United States of America

Contents

Preface

Amidst an onslaught of electronics dictionaries hitting the shelves of bookstores every year, here comes ***Electronics Terminology: A Concise Dictionary***. We at Informatik Rezurch believe that this is a wonderful resource for anyone working in or interfacing with the electronics world.

Although it is lightweight and concise, this dictionary manages to round up just about all the hottest acronyms and technical terminology encountered daily in electronics. Whether you are in engineering or marketing, acronyms like AID, CARs, MCM, and EDGE can be confusing and meaningless. Who needs to feel uninformed or self-conscious when someone is talking about ISO 9000, AOI, or sintering? With this pocket reference, you will always be able to quickly find the terms and buzzwords you need to keep up with the latest topics.

Our staff at Informatik Rezurch has spent many sleepless nights researching data from various current sources and compiling them in a user-friendly, easy-access format, which you can keep with you as you go. We acknowledge the importance of continuous product movement, and want to hear from you so that we can continue to improve this resource.

Please write to us if you discover any errors or believe that other materials could be added to make this little wonder more value-added. And for now, buckle your seat belt and enjoy the ride through this little wonder.

We thank you for your selection !

ACRONYMS

A

A: angstrom unit
ABC: activity-based cost
ABS: Acrylonitrite Butadiene Styrene
AC: AutoCAD certification/alternating current
ACPI: automated component placement and insertion
ACS: American Ceramics Society
ADC: analog-to-digital converter/ American Display Consortium
ADE: application development environment
ADP: automatic data processing
ADPCM: adaptive differential pulse code modulation
AEA: American Electronics Association
AEE: ambient electromagnetic environment
AES: Auger Electron Spectroscopy
AF: audio frequency
AFC: automatic frequency control
AFE: analog front end
AFM: atomic force microscopy
AGC: automatic gain control
AgPd: silver palladium
AI: automatic inspection/artificial intelligence
AID: automatic insertion DIP
AIN: aluminum nitride
Al2O3: aluminum oxide/alumina
ALT: accelerated life cycle test
ALU: arithmetic logic unit
AM: automated mapping/amplitude modulation
AMI: alternate-mark inversion
AMLCD: active matrix liquid crystal display
AMPS: advanced manufacturing and prototyping service

AMRAAM: advanced medium range air-to-air missile
AMRC: Assembly Marketing Research Council
ANA: automatic network analyzer
ANSI: American National Standards Institute
AP: arithmetic processor
APC: advanced process control/array processor controller
APC-7: standard precision 7-mm microwave coaxial connector
APE: asynchronous processing element
APG: algorithmic pattern generator
API: application programming interface/ asynchronous pixel interface
APIO: array processor input/output
APS: asynchronous packet switch
AQL: acceptable quality level
ARM: advanced RISC machines
ARPA: Advanced Research Projects Agency
ART: advanced receive & transmit
ARTE: advanced receive & transmit-enhanced
As: arsenic
ASM: American Society for Materials
ASME: American Society of Mechanical Engineers
ASP: addressable scan port/advanced signal processor
ASQC: American Society for Quality Control
ASSP: application specific standard products
ASTM: American Society for Testing and Materials
ASU: application-specific unit
ATAB: area-array tape-automatic bonding
ATAPI: AT attachment packet interface
ATE: automated test equipment
ATM: asynchronous transfer mode
ATPG: automatic test pattern generation

A2LA: American Association for Laboratory Accreditation
Au: gold
AuGe: gold-germanium
AuSi: gold-silicon
AuSn: gold-tin
AV: audiovisual
AVLP: automatic vehicle location and positioning system
AWG: American wire gauge/arbitrary waveform generator
AWP: advanced wafer passivation

B

B: boron
BCB: benzocyclobutene
BCCD: bulk channel charge-coupled device
BCDS: bulk chemical distribution systems
BCW: bare chip and wire
Be: beryllium
BeO: beryllia
BER: bit error
BIB: burn-in board
BICFET: bipolar inversion-channel field effect transistor
BICMOS: bipolar and complementary metal oxide semiconductor
BICMOS-II: 1.2 micron BICMOS
BICMOS-III: 1.0 micron BICMOS
BICMOS-IV: 0.8 micron BICMOS
BICMOS-V: 0.5 micron BICMOS
BIST: built-in self-test
BIT: bump interconnect technology
BIT-BLTs: bit-block boundary/logical transfers

BITE: built-in test equipment
BLB: beam lead bonding
BLDC: brushless DC (see DC)
BLM: ball-limiting metallurgy
BMC: bulk molding compound
BOD: biological oxygen demand
BOM: bill of material
BOPS: billion operations per second
BPR: business process re-engineering
BQFP: bumpered quad flat pack
BRT: bias resistor transistor
BSDL: boundary scan description language
BSE: back scattered electron
BSM: backside metallurgy
BSP: buffered serial port
BSR: boundary scan register
BT: bismaleimidetriazine resin
BTAB: bumped tape automated bonding
BTE: behind-the-ear
BTL: backplane transceiver logic
BTU: British thermal unit
BW: bandwidth
BWO: backward wave oscillator

C

C: carbon/centigrade/capacitor
CA: cellulose acetate
CAD: computer-aided design
CAE: computer-aided engineering
CAI: computer-aided instruction

CAID: computer-aided industrial design
CALS: computer-aided acquisition and logistics support
CAM: content-addressable memory/ computer-aided manufacturing
CARs: corrective action requests
CAS: column address strobe/computer-assisted specifying
CAT: computer-aided technology/ computer-aided testing
CAVP: complex arithmetic vector processor
C & W: chip and wire
CBA: chemical blowing agent
CBE: chemical beam epitaxy
CBGA: ceramic ball grid array
CBR: constant bit rate
CC: chip carrier
CCA: circuit card assembly/California Circuit Association
CCB: controlled-collapse bonding
CCC: ceramic chip carrier
CCD: certified companies directory/ charge-coupled device
CD: critical dimension
CDA: clean dry air
CDB: channel data buffer
CDIP: ceramic dual in-line package
CDPD: cellular digital packet data
CDR: critical design review
CDRAM: cache DRAM
CE: capillary electrophoresis
CEM: contract electronic manufacturing
CERDIP: ceramic dual in-line package
CERQUAD: ceramic quad flat pack
CFA: current-feedback amplifier
CFC: chlorofluorocarbon/chlorinated fluorocarbon

CFD: computational fluid dynamics
CGA: configurable gate array/column grid array
CGMP: current good manufacturing practices
CIC: copper invar copper
CIG: computer integrated geography
CIM: computer integrated manufacturing
CIS: component information system/card information structure
CISC: complex-instruction-set computing
CISPR: International Special Committee on Radio Interference
CIV: corona inception voltage
CL: total load capacitance
CLA: centerline average
CLAN: configurable local area network
CLB: configurable logic block
CLCC: ceramic leaded chip carrier
CLE: coefficient of linear expansion
CLGA: ceramic land grid array
CM: configuration management/centimeter
CMC: common mezzanine card/copper molybdenium copper
CMOS: complementary metal-oxide semiconductor
CMP: chemical mechanical polishing
CMRR: common-mode rejection ratio
CMT: chip-mounting technology
CNC: computer numerical control
COB: chip-on-board
COBRA: constant bit rate ATMs
COD: chemical oxygen demand
CODEC: code/decode
COF: chip-on-flex
COG: chip on glass/cost of good

COM: computer output microfilm
COO: cost of ownership
COP: crystal originated particles
COT: customer owned tooling
COTS: commercial off-the-shelf
Cp: capability index
CP: component placement/central processor
CPF: color pixel formatter
CPGA: ceramic pin grid array
CPH: components per hour
CPI: cycles per instruction
CPLD: complex programmable logic device
CPU: central processing unit
CQFP: ceramic quad flatpack
CQJB: ceramic quad J-bend
CQP: ceramic quad pack
CQUAD: ceramic quad package
CRADA: cooperative research & development agreement
CRC: cyclic redundancy check
CRIM: conventional reaction injection molding
CROM: control and read-only memory
CRT: cathode-ray tube
CSA: Canadian Standards Association
CSIC: customer-specified integrated circuits
CSP: chip scale package/common signal processing
CT: cordless telephone
CTE: coefficient of thermal expansion
CTFE: chlorotrifluoroethylene
CTR: current transfer ratio
CTTE: custom thermal test enclosure
CV: capacitance-voltage
CVD: chemical vapor deposition
CVSD: continuously variable slope delta

D

DAA: data access arrangement
DAC: digital-to-analog converter
DAM: data addressable memory
DASD: direct-access storage device
DAVID: digital audio/video interactive decoder
dB: decibel
DBS: direct broadcast system
DC: direct current
DCA: direct chip attach
DCAS: Defense Contracts Administration Services
DCB: direct copper bond
DCFL: direct coupled FET logic
DD: design development
DDE: dynamic data exchange
DDS: direct digital synthesis
DECT: digital European cordless telephone
DFA: design for assembly
DFF: design for fabrication
DFM: design for manufacture
DFMA: design for manufacture and assembly
DFSM: dry film solder mask
DFT: design for test
DFX: design for analysis
DGA: die grid array
DI: direct imaging/deionized water
DIF: digital image folding
DIMM: dual in-line memory module
DIP: dual in-line package

DLL: dynamic-link libraries
DLM: dual-layer metal
DMA: direct memory access

DMC: drive manager chip
DMM: digital multimeter
DMOS: double-diffused metal oxide silicon/double-diffused metal oxide semiconductor
DO: defect opportunity
DOD: Department of Defense
DOE: Department of Energy
DOF: depth of focus/ degree of freedom
DOS: disk-operating system
DPA: destructive physical analysis
DPR: dual port RAM
DPST: double pole single throw
DPU: data processor unit/defect per unit
DRC: design rule checking
DRO: dual read-out/dielectric resonator oscillator
DRPI: data representation procedural interface
DSC: differential scanning calorimetry
DSO: digital storage oscilloscope/dielectric stabilized oscillator
DSP: digital signal processor
DSSP: digital sensor signal processing
DSSS: direct sequence spread spectrum
DTA: differential thermal analysis
DTAB: demountable tape-automated bonding
DTL: diode transistor logic
DTMF: dual-tone multifrequency
DUT: device under test
DVDTSP: digital-value discrete-time signal processing
DVM: digital voltmeter

DWF: dice in wafer form
DXF: data exchange format

E

E: potential
EAC: European Accreditation of Certification
EAPROM: electrically alterable programmable read-only memory
EC: European Community
ECC: error correction code/error checking and correction
ECL: emitter-coupled logic
ECM: electronic countermeasure
ECO: engineering change order
ECP: enhanced capabilities port
ECTFE: ethylenechlorotrifluoroethylene
ED: electrodeposited
EDA: electronic design automation
EDAC: electronic design automation companies
EDFM: electronic design of manufacture conferencing
EDGE: electronic design guide for excellence
EDIF: electronic design interchange format
EDM: electrical discharge machine/ engineering data management
EDO: extended data out
EDO DRAM: extended data out DRAM
EDP: electronic data processing
EDRAM: enhanced DRAM
EDS: energy dispersive spectroscopy
EECA: European Electrical Component Manufacturers

Association
EEPROM: electrically erasable programmable read-only memory
EFO: electronic flame-off
EFTA: European Free Trade Association
EHF: extremely high frequency
EIA: Electronic Industries Association
EIAJ: Electronics Industry Association of Japan
EIF: Electronic Industries Foundation
EIPC: European Institute of Printed Circuits
EISA: enhanced ISA
ELF: extremely low frequency
ELR: electronic load relay
EM: electromagnetic
EMB: enhanced mask burst
EMC: electromagnetic compatibility
EMF: electromagnetic force
EMI: electromagnetic interference
EMP: electromagnetic potential
EMR: electromagnetic radiation/ electromechanical relay
EMSI: Electronic Manufacturing Services Industry
EM/SME: The Association for Electronics Manufacturing of the Society of Manufacturing Engineers
ENOB: effective number of bits
ENR: excess noise ratio
EOL: end of life
EOS: electrical overstress
EP: epoxy
EPA: Environmental Protection Agency
EPAC: electrically programmable analog circuit
EPDA: electronic product development automation
EPIC: environmentally protected integrated circuit
EPLD: erasable programmable logic device

EPP: enhanced parallel port
EPR: ethylene propylene rubber
EPROM: electrically programmable read-only memory
ESCA: electron spectroscopy for chemical analysis
ESD: electrostatic discharge
ESDA: electronic system design automation
ESDS: electrostatic discharge sensitive
ESS: environmental stress screening
ETACS: European Total Access Communication System
EUT: equipment under test
EVA: ethylene vinyl acetate
EXUS: execution unit

F

FA: failure analysis
FACI: first article configuration inspection
FAE: field application engineer
FAM: field application manager
FBT: functional board test
FCB: flip chip bonding
FDDI: fiber distributed data interface
FDDT: finite difference discrete time
FEA: finite element analysis
FED: field emission display
FET: field effect transistor
FFT: fast Fourier transform
FIB: focused ion beam
FIC: film integrated circuit
FIFO: first-in first-out

FIP: framed-in-place
FLCD: ferro electric liquid crystal display
FM: frequency modulation/facility management
FMEA: failure mode and effects analysis
FMECA: failure mode, effects, and criticality analysis
FO: fiber optics
FOC: flex-on-chip
FOGW: fiber-optic ground wire
FOV: field of view
FP: flat pack
FPC: fine pitch chip carrier/flexible printed circuit
FPCB: field programmable circuit board
FPD: flat panel display
FPIC: field programmable interconnect devices
FPGA: field-programmable gate array
FPT: fine pitch technology
FPU: floating point unit
FPW: flexible printed wiring
FQFP: fine-pitch quad flat pack
FR: failure rate
FRED: fast recovery epitaxial diode
FR-4: flame retardant epoxy glass laminate
FRP: fiber reinforced plastic
FSB: functional system block
FSDS: flat screen display system
FSK: frequency shift keying
FSM: finite state machine
FSR: full scale range
FTIR: Fourier transform infra-red
FTP: file transfer protocol
FTR: functional throughput rate
FTY: first-time yield

G

GaAs: gallium arsenide
GAC: granulated activated carbon
Gb: gigabyte
GCB: generic component board
GCG: gas-to-chemical generation
GDT: geometric dimensioning and tolerancing
Ge: germanium
GEO: geosynchronous orbit
GFAAS: graphite furnace atomic absorption spectroscopy
GHz: gigahertz
GII: global information infrastructure
GIS: geographic information system
GND: ground
GPIB: general purpose interface bus
GPL: graphical programming language
GPS: global positioning system/global positioning satellite
GPT: general-purpose timer
GSM: global system for mobile communications
GTAB: ground plane tape-automated bonding
GTO: gate turn-off
GUI: graphical user interface

H

HAL: hot air leveling
HALS: hot air leveled solderboard/ hardware abstraction layer

HASL: hot air solder leveled
HAST: highly accelerated stress test
HAZMAT: hazardous material
HCC: hermetic chip carrier
HCMOS: high density CMOS
HCSM: high conformance solder mask
HDA: head disk assembly/high density array
HDBGA: high density ball grid array
HDC: head disk controller
HDCM: high-density ceramic module
HDEA: high density embedded array
HDI: high density interconnect
HDL: hardware description language
HDMI: high density microelectronics interconnections
HDP: high density plasma
HDPE: high density polyethylene
HDSMT: high density surface mount technology
HDT: heat distortion temperature
HDTV: high definition television
HEMT: high electron mobility transistor/high energy mobility transistor
HES: hot electron stress
HF: high frequency
HFF: high frequency fundamental
HIC: hybrid integrated circuit
Hi-K: high dielectric constant
HIP: hex-in-line package
HLD: high level debugger
HMIC: hybrid microwave integrated circuit
HMOS: high-performance MOS
HPI: host processor interface/host port interface
HPLC: high-pressure liquid chromatography
HPMs: hazardous production materials/ high power

microwaves
HQFP: heat spreader quad flat pack
HSE: high-selectivity etch
HSPN: high speed plastic network
HSS: high-side switch
HTRB: high temperature reverse bias

I

IAC: instrument-on-a-card
IAPU: image array processing unit
IBIS: I/O buffer information specification
IC: ion chromatography/integrated circuit
ICCAD: International Conference on Computer-Aided Design
ICE: in-circuit emulator
ICP-MS: inductively coupled plasma mass spectrometry
ICR: interconnect ring
ICSPAT: International Conference on Signal Processing Applications and Technology
ICT: in-circuit test
IDC: insulation displacement connector
IDE: interactive development environment
IDT: integrated device technology
IEC: International Electrotechnical Commission
IEEE: Institute of Electrical and Electronics Engineers
IEEJ: Institute of Electrical Engineers of Japan
IEPS: International Electronics Packaging Society
IGBT: insulated gate bipolar transistor
IHS: information handling service

IIE: Institute of Industrial Engineers
ILB: inner lead bond
ILD: interlayer dielectric
ILO: injection-locked oscillator
IMB: intermodule bus
IMC: insertion mount component
IMD: intermetal dielectric/intermodulation distortion
IME: Institute for Microelectronics
IMPAT: impact ionization avalanche and charge transit time
IMS: insulated metal substrate
In: indium
InAs: indium arsenide
Internet: international network
I/O: input/output
IOC: I/O channel
IPC: Institute for Interconnecting and Packaging Electronic Circuits/integrated passive component/interconnecting and packaging electronic circuits
IPEC: integrated passive electronic components
IPS: instructions per second
IR: infrared
IrDA: infrared data association
IRED: IR emitting diode
IRS: infrared scan
ISA: instruction set architecture/industry standard architecture
ISDN: integrated services digital network
ISHM: International Society for Hybrid Microelectronics
ISM: industrial, scientific and medical
I/SMT: interconnect/surface mount technology
ISO: International Standards Organization
ISO-MPEG: International Standards Organization Motion

Picture Experts Group
ISR: integrated switching regulator
ITC: International Test Conference
ITE: in-the-ear
ITRI: Interconnection Technology Research Institute
IX: ion exchange

J

JC: JEDEC committee
JEDEC: Joint Electronic Devices Engineering Council
JEIDA: Japanese Electronic Industry Development Association
JESSI: Joint European Submicron Silicon Initiative
JFET: junction field-effect transistor
JIT: just-in-time
JPEG: Joint Photographic Experts Group
JTAG: Joint Test Action Group

K

K: dielectric constant
KGD: known good die
KME: Kyushu Matsushita Electric
KSIA: Korean Semiconductor Industry Association

L

LAB: logic array block
LAN: local area network
LBA: loaded board analyzer
LBO: line buildout
LC: inductance-capacitance
LCC: leadless chip carrier
LCCC: leadless chip carrier connection
LCD: liquid crystal display
LCP: liquid crystal polymer
LDCC: leaded chip carrier
LDPE: low-density polyethylene
LE: logic element
LEAF: law enforcement access field
LED: light-emitting diode
LEO: low earth orbit
LFSR: linear feedback shift register
LGA: land grid array
LIC: linear integrated circuit
LICA: low inductance capacitor array
LID: leadless inverted device
LIF: low insertion force
LIFO: last in first out
LIM: liquid injection molding
LISP: list processor
LITE: low impact testability enhancement
LLCC: leadless chip carrier
LMCH: leadless multiple chip hybrid
LPCVD: low-pressure chemical vapor deposition
LPGA: laser programmable gate array
LPISM: liquid photoimageable solder mask
LPM: Library of Parameterized Modules

LOC: leads on chip
LOCOS: local oxidation of silicon
LPCVD: low pressure chemical vapor deposition
LPE: liquid phase epitaxy
LPGA: leadless pad grid array
LPISM: liquid photoimageable solder mask
LRM: language reference manual/line replaceable module
LSA: limited space-charge accumulation
LSB: least significant bit
LSI: large scale integration
LSS: low-side switch
LSSD: level-sensitive scan design
LTPD: lot tolerance percent defective
LUT: lookup table
LVA: low-voltage avalanche
LVBO: low voltage battery operated
LVTTL: low voltage TTL (see TTL)

M

MAC: media access controller
MANTECH: manufacturing technology program
MAU: medium attachment unit
MBE: molecular beam epitaxy
mBGA: mini ball grid array
Mbps: megabits per second
MBT: metal base transistor
MC: metallized ceramic
MCAD: mechanical CAD (see CAD)
MCC: Microelectronics and Computer Technology Corp/ miniature chip carrier

MCCI: multichannel communication interface
MCM: multichip module
MCM-C: ceramic MCM
MCM-D: deposited MCM
MCM-L: laminate MCM
MCM-L/O: MCM laminate/overlay
MCP: multichip packages or also few-chip packages.
MCR: molded carrier ring
MCRPQFP: molded carrier ring plastic quad flat pack.
MCU: microcontroller unit
MDA: mechanical design analysis
MDIP: molded dual in-line package
MEK: methyl ethyl ketone
MELF: metal electrode face
MEMS: microelectromechanical system
MEPCB: molded encapsulated printed circuit board
MEPPE: Microelectronic Packaging and Processing Engineers
MESFET: metal-semiconductor field-effect transistor
MF: microfiltration
MFC: mass flow controller
MFD: microfarad/microelectronic functional device
MFS: minimum feature size
MHP: multichip hybrid package
MIC: microwave integrated circuit
MII: media independent interface
MIL-STD: military standard
MIPS: millions of instructions per second
MIS: metal insulator semiconductor
MLB: multilayer board
MLC: multilayer ceramic capacitor
MLM: multi-level metal

MMCM: multi-modular computer module
MMG: Multibus Manufacturers Group
MMIC: monolithic microwave integrated circuit
MMPQFP: multilayer molded plastic quad flat pack
MMU: memory management unit
mmW: millimeter wave
MNOS: metal nitride-oxide semiconductor
MNP: microcom networking system
MNS: metal nitride semiconductor
MO: magneto-optical
MOA: Ministry of Aerospace
MOPS: million operations per second
MOS: metal-oxide semiconductor
MOSFET: metal-oxide semiconductor field-effect transistor
MOU: memorandum of understanding
MOV: metal oxide varistor
MOVPE: metalorganic vapor phase epitaxy
MPAC: mask programmable analog circuit
MPEG: Motion Picture Experts Group
MPP: multimedia playback processor
MPU: microprocessing unit
MQFP: metric quad flat pack
MRB: Material Review Board
MS: microstrip
MSB: most significant bit
MSI: medium scale integration
MSL: microstrip planar asymmetric transmission system
MSM: mass storage module
MSO: mixed signal option
MSPS: megasamples per second
MTBF: mean time between failure
MTH: metallized through-hole

MTM: multiple termination module
MTPS: mega-transfers per second
MTTF: mean time to failure
MTTR: mean time to repair
MUX: multiplexer
MVA: motion video architecture
MVP: multimedia video processor
MVT: moisture vapor transmission
MWPS: million words per second

N

NAFTA: North American Free Trade Agreement
NATA: North American Telecommunications Association
NC: numerically controlled/normally closed
NCdrill: numerically controlled drill
NCP: no-clean process
NDE: non-destructive evaluation
NDRO: non-destructive readout
NDT: non-destructive test
NEDA: National Electronic & Distributors Association
NEMA: National Electrical Manufacturers Association
NF: nanofiltration
NFF: no form feed
NGA: noise gain analyzer
Nibble: 4 bits
NiCd: nickel-cadmium
NICET: National Institute of Certification for Engineering Technologies
NII: National Information Infrastructure
Ni-MH: nickel-metal hydride, a rechargeable battery

NIRA: near infrared reflectance analysis
NIST: National Institute of Standards & Technology
NMOS: N-channel metal-oxide silicon
NOP: number of points
NOS: network operating system
NPO: negative and positive zero
NPT: non-punch-through
NRC: National Research Council
NRE: nonrecurring engineering
NSSN: National Standards Systems Network
NST: Nyquist Sampling Theorem
NT: no TAB/network termination
NTDS: Navy Tactical Data Systems
NTFS: NT's file system
NTIS: National Technical Information Service
NTSC: National Television Standards Committee

O

OA: organic acid
OCL: open-circuit inductance
OCR: optical character recognition
OCV: optical character verification
OCXO: oven-controlled crystal oscillator
ODBMS: object database management system
ODP: ozone depletion potential
ODS: ozone-depleting substance
OEIC: optoelectronic integrated circuit
OEM: original equipment manufacturer
OFHC: oxygen-free high-conductivity
OHDL: OrCAD hardware design language
OLB: outer lead bond

OLE: object linking and embedding
OLTP: on-line transaction processing
OMC: organic matrix composite
OMF: open model forum
OMPAC: overmolded plastic pad array carrier
OPP: optical precipitate profiler
OPS: operations per second
OSCG: on-site chemical generation
OSF: open systems foundation/open software foundation
OSHA: Occupational Safety and Health Act
OSP: organic solderability preservative
OTP: one-time programmable memory

P

PAA: pad area array
PAD: programmable address decoder
PAL: programmable array logic/phase alternation line
PBGA: plastic ball grid array
PC: personal computer
PCB: printed circuit board
PCBT: pressure cooker bias test
PCI: peripheral component interconnect
PCMCIA: Personal Computer Memory Card International Association
PC-QFP: printed circuit based quad flat pack
PCR: plastic chip carrier
PCT: pressure cooker test

PCTFE: polychlorotrifluorethylene
PCU: peripheral control unit
PCWG: Personal Conference Work Group
PDA: personal data assistant/percent defective allowable
PdAg: palladium silver
PDEF: physical design exchange format
PDIF: P-CAD data interchange format
PDIP: plastic dual in-line package
PDM: product data management
PECVD: plasma-enhanced chemical vapor deposition
PEL: picture element in display
PF: power factor
PFC: power factor correction/perfluorocarbon
P4: photolithographic pattern plated probe
PGA: pad or pin grid array
PHP: power hybrid package
PHR: parts per hundred parts resin
P/I: packaging and interconnecting
PIC: photonic integrated circuit/personal intelligent communications/power integrated circuit
PICMG: PCI Industrial Computer Manufacturer Group
PIN: p-n junction with isolation region
PIND: particle impact noise detection
PIP: pin insertion package
Pixel: picture element
PLA: programmable logic array
PLCC: plastic leaded chip carrier
PLD: programmable logic device
PLDCC: plastic leaded chip carrier
PLGA: plastic land grid array
PLI: programming language interface
PLL: phase lock loop
PMOS: P-channel metal-oxide semiconductor

PMA: post-metallization anneal
PMC: PCI mezzanine card
PMCM: plastic multichip module
PMP: premolded plastic
PMU: parametric measurement unit
PN: processor node
POS: porcelain-on-steel
POST: power-on self tests
POUCG: point of use chemical generation
PPGA: plastic pin grid array
PPM: parts per million
PPP: parallel pipeline processor/point-to-point protocol
PPT: precision pad technology
PQFP: plastic quad flat pack
PROM: programmable read-only memory
PRS: pattern recognition system
PSCR: photosensitive copper reduction process
PSD: power spectral density
PSG: phospho-silicate glass
PSI: pounds per square inch
PSOP: power small outline package
PSP: programmable signal processor
PSRAM: pseudo-static RAM
PSTN: public switched telephone network
PtAu: platinum gold
PTF: polymer-thick-film
PTFE: polytetrafluoroethylene
PTH: plated through hole/pin through hole
PTV: projection television
PUR: polyurethane
PVD: peak voltage detect/physical vapor deposition
PWA: printed wiring assembly
PWB: printed wiring board

PWM: pulse width modulator

Q

QCI: qualification conformance inspection
QFB: quad flat butt-leaded package
QFP: quad flat pack
QPL: qualified products list
QSM: queued serial module
QUIP: quad in-line package

R

RA: rosin activated flux
RAB: Registrar Accreditation Board
RAID: redundant arrays of inexpensive disks
RAM: random access memory

RAMDAC: random-access memory/digital-to-analog converter
RAMTF: random access matrix tray feeder
RAS: row address strobe
RCC: rectangular chip carrier
RDBMS: relational database management system
RDRAM: RAMBUS DRAM
RF: radio frequency
RFI: radio frequency interference
RGA: residual gas analysis
RGB: red-green-blue
RH: relative humidity

RI: receiving inspection
RIE: reactive ion etching
RISP: rapid impingement speed plating
RMA: rosin mildly activated
RMI: resource manager interface/ repetitive-motion injury
rms: root mean square
RO: reverse osmosis
ROI: region of interest/return on investment
ROM: read-only memory
ROW: rest of the world
RP: reinforced plastic/rapid prototyping
RPM: revolutions per minute
RSSI: received signal strength indicator
R3: reduced radius removal
RTL: register transfer level/resistor-transistor logic
RTM: resin transfer molding
RTOS: real-time operating system
RTP: rapid thermal processing/reinforced thermoplastic

S

SACVD: subatmospheric chemical vapor deposition
SAM: serial access memory/scanning acoustic microscope/scanning Auger microprobe
SAMPE: Society for the Advancement of Material and Process Engineering
SAW: surface acoustic wave
SBC: surround by complement/single board computer
SBDIP: side-brazed dual in-line package
SBIR: Small Business Innovation Research

SCC: stackable ceramic carrier/Standards Council of Canada/square chip carrier/stress corrosion cracking
SCD: specification control drawing
SCE: storage control element
SCFL: source coupled FET logic
SCIM: single chip integration module
SCM: single chip module
SCOT: sealed chips on tape
SCP: single chip package
SCPI: standard commands for programmable instruments
SCSI: small computer system interface
SCXI: signal conditioning extension for instrumentation
SDBGA: super dissipation ball grid array
SDCM: serial data converter modules
SDF: standard delay file/standard delay format
SDH: synchronous digital hierarchy
SDK: system development kit
SDSC: Standards & Data Services Committee
SECS: Semiconductor Equipment & Material International's Equipment Communications Standard
SEM: scanning electron microscope
SEP: someone else's problem
SEU: single event upset
SFI: solder free interconnections
SFP: short-fault protection
SFROMÔ: smart flash ROM
SGA: solder grid array or BGA (see BGA)
SHF: super high frequency
Si: silicon
SIA: Semiconductor Industry Association
SiC: silicon carbide
SICL: standard instrument control library
SIG: special interest group

SIM: system integration module/single in-line module
SIMD: single instruction multiple data
SIMM: single-in-line memory module
SIMOX: separation by implanted oxygen
SIMS: secondary ion mass spectrometry
SiN: silicon nitride
SIP: single-in-line package
SIR: surface insulation resistance
SIU: system interface unit/system integration unit
SLAM: single-layer alumina metallized
SLC: surface laminar circuit/single layer ceramic
SLD: source level debugger
SLICC: slightly larger than IC carrier
SLT: solid logic technology
SMA: surface-mounted assembly/standard miniature microwave coaxial connector
SMART: surface mount and reflow technology
SMC: surface-mount components/S-Bus mezzanine card
SMD: surface-mount device
SMEMA: Surface Mount Equipment Manufacturers Association
SMI: small and medium sized industries
SMM: system management mode
SMOBC: solder mask over bare copper
SMP: surface-mount package
SMPGA: surface-mount pin grid array
SMT: surface mount technology
SMTA: Surface Mount Technology Association
SMU: source-measure unit
SO: small outline
SOB: small outline butt-leaded package
SOG: spin-on glass
SOI: silicon-on-insulator

SOIC: small-outline integrated circuit
SOJ: small-outline J-leaded package
SOLA: short, open, load and air-capacitor
SONET: synchronous optical network
SOP: small outline package
SOS: silicon on sapphire
SOT: small-outline transistor
SP: span processor/signal processor
SPARCÔ: scaleable processor architecture
SPC: statistical process control
SPDIP: shrink plastic dual in-line package
SPDT: single pole double throw
SPE: Society of Plastic Engineers
SPEC: Standard Performance Evaluation Corporation
SPI: serial-peripheral interface/synchronous pixel interface
SPICE: simulation program for integrated circuits emphasis
SPST: single pole single throw
SQC: statistical quality control
SQFP: shrink quad flat pack
SRAM: static random-access memory
SRC: Semiconductor Research Corporation
SRT: source routing transparent
SRU: smallest replaceable unit
SSD: solid-state disk
SSH: simultaneous sample and hold
SSI: small-scale integration
SSMM: solid-state mass memory
SSOP: shrink small outline package
SSR: solid state relay
SST: solid state technology
SSWS: static-safe work station

STL: simulation and test language/stripline
STM: scanning tunneling microscope
SVGA: super video graphic adapter
SVP: surge voltage protector
SWIFT: software interface technology

T

TA: Triazine A
TAAF: test analyze and fix
TAB: tape automated bonding
TAC: Technical Activity Committee
TAP: test access port
TAPI: telephone application program interface
T-BGAs: tape ball grid array
TC: thermocompression
TCC: temperature coefficient of capacitance
TCE: temperature coefficient of expansion
TCM: total chemical management/thermal conduction module
TCP: tape carrier package
TCP/IP: transmission control protocol/internet protocol
TCR: temperature coefficient of resistance/tape carrier ring
TDMAT: tetrakisdimethylamino titanium
TDR: time domain reflectometer
TDS: total dissolved solid
TE: transverse electric
TEC: thermal electrical cooler
TEGFET: two-dimensional electron-gas field effect transistor

TEM: transverse electromagnetic/transmission electron microscopy
TEOS: tetraethyl orthosilicate
TF: trim and form
TFT: thin-film transistor
Tg: glass transition temperature
TGA: thermal gravimetric analysis
TH: through-hole mount/track & hold
THB: temperature humidity bias
TIC: tape-automated bond in cap
TiN: titanium nitride
TIR: total indicated reading
TLB: translation lookaside buffer
TLC: thin-layer chromatography
TLIM: thermal limit
TLM: transmission line model/transmission line matrix/ triple-layer metal
TLP: two layer photo
TMA: thermo-mechanical analysis
TMRC: Technology Marketing & Research Council
TO: transistor outline package
TOC: total organic compound/total oxidizable carbon
TOI: third-order intercept
TOP: thin outline package
Tp: transition temperature
TPA: terminal position assurance
TPE: thermalplastic elastomer
TPS: AU power supply
TPU: time processor unit
TQFP: thin quad flat pack/tape quad flat pack
TQM: total quality management
Tr: transition time
TRAPATT: trapped plasma avalanche-triggered transit

TREE: transient radiation effect on electronics
T/R: transmit and receive
TRL: through-reflect-line
TRP: Technology Reinvestment Program
TS: thermosonic
TSM: technical sales manager/ thermosetting material
TSOP: thin small outline package
TSP: temperature-sensitive parameter
TSSOP: thin shrink small outline package/ thin scaled small outline package
TTL: transistor-transistor logic
TVI: TV interference
TVS: transient voltage suppresser

U

UBS: uninterruptible battery system
UCOM: Users' Committee of Foreign Semiconductors
UCS: user coordinate system
UDPs: user-defined primitives
UF: ultra filtration
UFP: ultra fine pitch
UFPT: ultra fine pitch technology
UHF: ultra high frequency
UHV: ultra high vacuum
UL: Underwriters Laboratory
ULSI: ultra large scale integration.
UNEP: United Nations Environment Program
UPC: universal product code
UPS: uninterruptible power source
US: ultrasonic

UTP: unshielded twisted pair
UTQFP: ultra-thin quad flat pack
UTSOP: ultra thin small outline package
UUT: unit under test
UV: ultra violet

V

VAR: value-added reseller
VBX: Visual Basic control
VC: venture capitalist
VCD: variable center distance
VCOS: visible caching operating system
VCXO: voltage-controlled crystal oscillator
VDE: visual development environment
VESA: Video Electronics Standards Association
VGA: video graphics array
VHF: very high frequency
VHSIC: very high speed IC
VIL: vertical in-line
VITA: VMEbus International Trade Association
VL Bus: VESA local bus
VLS: virtual library system
VLSI: very large scale integration
V-MOS: v-groove MOS
VNA: vector network analyzer
VOC: volatile organic compounds
VP: video processor
VPD: vapor phase dissolution
VPE: vapor phase epitaxy

VPS: vapor phase soldering
VQFP: very fine pitch quad flat pack
VR: virtual reality
VRAM: video random-access memory
VSAT: very small aperture terminal
VSIM: virtual surface image memory
VSO: very small outline
VSOP: very small outline package
VSS: VHDL system simulator
VXIBUS: VMEBUS extensions for instrumentation

W

WAN: wide area network
WB: wire bonded
WID: within-the-die
WIP: work-in-progress
WIW: within-the-wafer
WLAN: wireless local area network
WRAM: window RAM
WSI: wafer scale integration
WSTS: world semiconductor trade statistics
WUT: wafer under test
WWW: world wide web of the Internet
WYSIWYG: what-you-see-is-what-you-get

X

XIP: execute-in-place
X-MOS: high-speed metal oxide semiconductor

Y

YAG: yttrium-aluminum garnet
YTD: year to date

Z

ZIF: zero insertion force
ZIP: zigzag-in-line package
ZnO: zinc oxide

GLOSSARY

A

Abrasive Trimming: using abrasives to adjust resistance of resistors by cutting them.

Absolute Zero: the temperature where there is no molecular movement and heat dissipation in a material. Zero K = -273.15°C.

Absorption: the amount of moisture retained by a material.

Accelerated Aging: a process where voltage, temperature, mechanical and other parameters are taken to a point far exceeding their normal operating values so that the material under test can demonstrate destructive results in a short amount of time.

Acceptance Tests: tests required by the purchaser and performed by the vendor to determine the acceptability of the vendor's products.

Activating: a process where non-conductive materials undergo electroless plating.

Active Area: the actual usable substrate area. Normally, metal substrates tend to have the largest active area comparing to glass or ceramic substrates.

Active Devices: pertains to diodes, transistors & integrated circuits.

Active Metal: a metal such as iron that is extremely reactive and has a high electromotive force.

Active Network: contains both passive and active components.

Active Silicon Area (ASi): one of the main elements needed to determine the substrate efficiency. The substrate efficiency is determined by the ratio between Active silicon area and total substrate area.

Active Substrate: a substrate where both active and passive devices can be formed to provide discrete or integrated components.

Active Trimming: adjusting the circuit function by trimming certain resistors while the power is on.

Additive: a substance which, when added to materials, is capable of improving their properties.

Additive Process: a process where a conductive material such as copper or gold is selectively deposited onto the surface of a substrate to form conductor patterns.

Admittance: the contrary of impedance.

Additive Patterning: patterns are formed on a substrate by depositing resistive, conductor and dielectric materials sequentially onto it through a mask.

Adhesion: used to measure the bonding strength between two surfaces. It can be applied to two film layers sticking together.

Adhesion Layer: a layer used to support the thin metal layer in MCM-D technology. Chromium is typically used with copper to give the copper layer a better mechanical support.

Advanced Support System for Emulation and Test (ASSET): a testability and diagnostic system for designers to use boundary-scan test techniques to debug and verify operation of boards, ICs and systems.

Aging: a process where a material undergoes property changes with time when certain conditions are applied to it.

Agitation: a process where a plating bath is stirred

and excited so that more metal ions can be filled up at the cathode end to increase the deposition rate.

Alloy: two or more metals blended together in a solid-state solution.

Alumina: a popular substrate material for many MCM vendors in the design and development of co-fired and multilayer ceramics. It is also known as Alumina Oxide (Al_2O_3).

Aluminum: atomic no. 13 metal which is light and has a silvery white color.

Ambient Temperature: the atmospheric temperature in close contact with the electrical component.

American Bureau of Shipping (ABS): it is one of the major advocates of quality assurance and it offers certification to ISO9000.

Angle Connector: a connector for joining two conductors end to end at a certain angle.

Angle of Attack: the angle between the face of the squeegee and the surface of the screen in screen printing.

Angstroms: unit in thin film applications. One angstrom is roughly 0.0004 micron.

Anisotropic: the optical, thermal or electrical properties inside a material are different in different directions across the material.

Annealing: a way of strengthening certain metals by raising their body temperatures and cooling them slowly to make them less brittle.

Annular Ring: a ring-shaped conductor pattern that encircles a hole in a PCB.

Anode: the positive end in plating equipment where negative ions exit from the plating solutions.

Anodic Silver: a fine metal used in plating.

Anodization: a process where the thin-film resistor values can be changed or capacitor dielectric can be prepared through an electrochemical reaction. It is also for treating the metal surface with a protective layer that is non conductive.

Anti-Pad: a negative pad used to prevent any pin from connecting to an embedded metal layer.

Antioxidant: a substance capable of stopping or slowing down the oxygen decomposition of a material.

Application Specific Integrated Circuit (ASIC): a custom made circuit used for a specific application. It is typically manufactured in smaller quantities with a higher NRE cost.

Aqueous Cleaning: a cleaning treatment of electronic assemblies using de-ionized water or water-soluble agents.

Arc Resistance: the time necessary for an arc to build up conduction in a material.

Array: row or column arrangement of circuit components on a substrate.

Area Array TAB (Tape Automated Bonding): used in very complicated circuits such as VLSI circuits where pads on the edges and additional pads on the inner surface area of a chip are addressed in bonding.

As-Fired: the thick-film resistance values or the smoothness of a substrate surface immediately after firing in the furnace and before resistor trimming and surface polishing.

Aspect Ratio: the ratio between the film resistor's length and its width or the ratio between the length of a substrate and its width. For PCB, it is the ratio between the PCB thickness and the drilled hole diameter.

Asperities: the roughness and irregularities of surface.

A Stage: the beginning stage where a thermosetting resin is uncured.

Attach: the joining between components and their next level of interconnect (i.e. PWB or ceramic substrate).

Attenuation: the degradation of power level in an electrical system.

Au/Ge: a gold/germanium alloy with a 88 to 12 ratio that is suitable for high temperature die attachments.

Au/Sn: a gold/tin solder alloy with a 80 to 20 ratio that is suitable for gold-plated package sealing.

Autoclave: a closed tube used exclusively for conducting chemical reaction and other types of operation when high pressure and heat are applied to it.

Automatic Laser Trimming: an automatic process where thick film resistor patterns are cut by a laser to increase their values.

Automatic Optical Inspection (AOI): to inspect pattern or object using a camera in an automated system. The throughput rate is square inches inspected per second and the resolution is 2X the pixel size.

Automatic Wafer Sawing: a wafer is sliced into many little dice using an automated dicing or sawing machine.The machine is controlled via programs and operates at a high throughput rate.

Automatic Wire Bonder: a high throughput machine with a built-in microprocessor capable of performing fine wire bonding of dice. All critical parameters such as bond location, die height, bonding force and time, wire loop shape and height and etc are preprogrammed for accuracy reasons.

Average Chip Pitch: the average distance of separation between chip centers.

Average Interconnection Density: the ratio of the no. of signal layers to the average wire pitch.

Average Wire Length: the average of all the wire lengths of a package.

Axial Displacement: a shift in the position when a force is applied.

Axial Lead: leads emanating along the central axis from the ends of a discrete component.

B

Back Bonding: a process where integrated circuits are mounted on the substrate by applying adhesive to the back side of the devices.

Back End of the Link (BEOL): a process where a wafer is being sliced into many small integrated circuits, transistors and resistors are wire bonded to the substrate pads and devices are connected to the package.

Backplane Panel: a panel where printed wiring assemblies and integrated circuit packages can be plugged in and interconnected together.

Bake-Out: a high temperature cycling process where unwanted gases and moisture are permanently removed from a package before it is sealed.

Ball Grid Array (BGA): it is the latest surface mountable semiconductor package containing only solder balls as its leads. When undergoing solder reflow process onto the next level of interconnect (i.e. PWB), it yields no coplanarity problem. It offers I/Os from 20 to over 500 and has high assembly yield. Currently, there exists two major types - plastic BGA (PBGA) and ceramic BGA (CBGA). The overall cost is high, and as the level of market acceptance increases, it will come down eventually.

Bandwidth: a device's transmission capability as measured by the number of bits per second.

Betascope: an instrument relies on radioactive isotopes to measure film and plating thickness.

Bias Moisture Test: a process where an electrical bias is applied to a package inside a steam chamber to purposely accelerate the penetration of moisture into the package. Enough moisture will cause the gold wires and bonding pads inside the package to corrode.

Binders: an organic or inorganic material which, when added to thick film compositions, is capable of providing strengths for prefire handling.

Black Body: a body capable of radiating and absorbing radiant energy.

Blind Via: a via connecting one or more inner layers to the surface of a substrate.

Blisters: a phenomenon where layers of a laminate substrate are separating due to swelling. This could be applied to any laminate layers or a laminate layer and a metal clad such as copper.

Blocking: a current resisting ability of a semiconductor device due to its high internal resistance.

Blowhole: a hole found in a solder connection or joint due to outgassing during soldering.

Bond Envelope: a range where acceptable bonding

can be obtained using existing bonding parameters.

Bonding: the joining of components to achieve electrical, mechanical or sealing purposes.

Bond Lift-Off: the separation of the bonded lead from its corresponding bond pad.

Bond Pull Test: a test for measuring the bonding strength of a lead or wire.

Braze: a high temperature (above 800°F) fusion of metals with a nonferrous alloy.

Breakdown Voltage: the maximum threshold voltage level that a device is capable of handling.

B-Stage: a curing stage applied to thermosetting resin. Typically, there are three stages - A, B and C, and B stage is the intermediate stage.

Bulk Resistance: portion of the contact resistance used for cross-section, length, and material type.

Buried Capacitance (BC): a new way of manufacturing boards where distributed decoupling capacitance can be obtained by embedding thin dielectric layers within the board.

Buried Channel: a transfer channel underneath a semiconductor surface.

Buried Resistors: a method where thick film resistors are embedded within the inner layers of a multilayer substrate to reduce conductor lengths.

Buried Via: a via used to connect between inner layers of a substrate only.

Burn-In: a process where components are stressed electrically at an elevated temperature and over a sufficient duration of time. Marginally working components will fail during this process and be screened out.

Burning Rate: plastic materials burning at certain temperatures.

Burst Mode: a multiplexer channel used exclusively for an I/O device, hence, allowing the device to transmit at a higher rate.

Bus: a conductor capable of transmitting signal, power and ground from one point to another point.

Bus Bar: a heavy duty metal bar made of either copper or aluminum that is capable of transmitting a large amount of current.

Bussing: the linking of two or more circuits.

Butt: to connect two conductors end to end and with their axis in line so that there is no overlap.

Butt Connector: a connector capable of doing butting (see "Butt").

Butt Joint: a connection between the PWB (printed wiring board) land pattern and the end of a component lead.

Butt Splice: an instrument capable of performing butting (see "Butt").

Butt Welding: a process where two metals are joined together in the absence of overlapping and brazing.

Byte: a computer storage unit with eight bits.

C

Cache Hit: cache memory containing the needed information.

Cache Memory: a high performance memory which is physically situated between the processor and its local memory.

Cache Miss: cache memory not containing the needed information.

Camber: the amount of warpage in a substrate or lead

frame.

Capacitance Density: the amount of capacitance per unit area. The unit is either pF/mil^2 or F/in^2.

Capacitor: a device capable of delivering any current for the purpose of maintaining a constant voltage.

Capillary: a bonding tool necessary for guiding a bonding wire inside its hollow tube and applying pressure to the wire during bonding.

Capillary Action: a liquid is drawn into a tiny opening due to surface tension effect.

Capillary Tool: a wire bonding tool used to feed the wire to its bonding surface through a bore along its long axis.

Castellation: the groove on a side of any leadless chip carrier used for electrical connection.

Catalyst: a substance capable of influencing the chemical reaction rate without altering itself at the end of the reaction.

Cavity: a package bottom with a recessed or an open area, or an area on the substrate where components were to be mounted.

Centerline Average (CLA): an imaginary mean centerline drawn between peaks and valleys on any given surface finish. Actual measurements are performed on the surface. The mathematical average of deviations between measured and imaginary numbers yields the CLA.

Centerwire Break: an event involving wire fractures at midspan during the wire pull test.

Ceramic: materials such as alumina, beryllia, steatite and other inorganic and nonmetallic types whose final characteristics will be decided after being fired at high temperatures. These materials are commonly used in various areas in microelectronics, with substrate and package being the most frequent applications.

Ceramic Board: a ceramic substrate where conductor patterns are printed to facilitate interconnects between circuit elements.

Ceramic package: typically, alumina is the material of choice over beryllia due to its ease of handling and cost advantage. A surface mountable cerdip-like package with lead frames on all four sides. Other names such as cerquad and cerpak are also used.

Ceramet: a metal flat pack having a ceramic bottom construction.

Cerdip: a ceramic dual in-line package containing a lead frame sandwiched between two layers and sealed when a glass frit is fired.

Cermet: a well-balanced solid material consisting of a delicate mixture of metal and ceramic substance in close form.

Characteristic Impedance: the ratio of the square root of the series inductance L to the square root of the shunt capacitance C in a transmission line.

Chemical Vapor Deposition: a chemical vapor reduction method of depositing circuit elements onto a substrate.

Chip: a device capable of being active or passive, discrete or integrated, and available in either packaged or bare form.

Chip-and-Wire: faced-up chip devices are wire bonded onto the substrate.

Chip Aspect Ratio: the ratio between the length and width of a chip resistor.

Chip Carrier: a surface mount package that is low profile and normally square. The chip mounting area or cavity occupies a large portion of the package and there are external connections in lead or pad form on all four sides of the package.

Chips-First: a different way of mounting chips onto the substrate. Chips are placed inside a milled cavity in a ceramic substrate, then the interconnect structure consisting of conductor layers and thin film dielectric is fabricated over the chips. Another name is Overlay Interconnect Technology.

Chip-on-Board (COB): an assembly technology which bonds bare chips directly onto printed wiring board using die attach adhesive and wire bonding technique. The chip and wires are then encapsulated using epoxy glob-top encapsulant for environmental protection.

Chip-on-Flex (COF): an assembly technology where bare chips are directly bonded onto flex circuit board. The wires are either gold or aluminum and the wire diameter is typically 1 or 1.25 mils.

Chip Scale Package (CSP): it is the latest and smallest package in the packaging world today. Its size is slightly larger than the die it contains. A low profile is made possible due to a thin plastic resin coating over the chip. Electrical contact with the PCB is made through the electrode bumps at the bottom of the device.

Chipshooter: a fast handler and placer of components having the same value and size.

Chuck: a part of the bonding machine used for holding the UUB (unit under bond).

Clad: a material bonded with a thin layer of metal on one or both sides.

Clock Skew: the maximum time difference between the earliest and the latest clock pulses arriving at the fanouts or sink pins in a circuit.

Coefficient of Thermal Expansion (CTE): a ratio between change in dimension and change in temperature as expressed in the form of DD/DT.

Co-fired: a process which prints both thick film conductor and resistor paste onto a green (raw) ceramic and fires them in a furnace simultaneously. The firing temperature will determine if it is a high temperature or low temperature co-fired process. Typically, 1600°C is for high temperature and 850°C for low temperature.

Column Address Strobe (CAS): it is used to drive the CAS input of either a VRAM (video RAM) or DRAM (dynamic RAM).

Concurrent Engineering: a systematic way of integrating all related functions necessary for the design and development of products at the beginning of the product life cycle. By implementing such an approach, engineers from design, testing and manufacturing, purchasing agents, marketers, and administrative staff can all work together to make the products a first-time success.

Conductivity: the electrical conduction of materials.

Conductor: metallic materials such as gold, copper, and silver that are low in resistivity (typically less than 0.0001W per cm) and capable of conducting electricity freely.

Conductor Adhesive: metal powders are added to an adhesive material to make it more conductive.

Conductor Epoxy: metal powders such as gold or silver are added to an epoxy material such as polymer resin to make it more conductive.

Conductor Spacing: the distance between adjacent

conductor lines printed or etched on the substrate surface or layers embedded inside a substrate.

Conductor Width: the width of each conductor line printed or etched on the substrate surface or layers embedded inside a substrate.

Conformal Coating: a coating material that is non-conductive and thin, and is frequently applied over electrical circuits for environmental and mechanical protection.

Contact Angle: an angle created between the bonding pad and bonding material.

Contact Printing: a screen printing technique where the screen is within a few mils away from the substrate. It is popular in printing using metal mask.

Contamination: the solderability limitation of any component footprint pad or soldering surface due to any uninvited particles or dust.

Controlled Collapse Bonding (CCB): during a flip-chip soldering process, controls the solder ball height reduction.

Controlled Collapse Chip Carrier Connection (C5): an assembly process used for depositing solder bumps onto the metallized pads on the underside of a pad array carrier and on the corresponding pads on the printed wiring board. The pad array carrier is then placed onto the board and reflowed to make electrical and mechanical connection.

Controlled Collapse Chip Connection (C4): also known as flip chip bonding. Bare chips are attached face-down to metallized substrates by aligning the I/O pads on the chips with the corresponding pads on the substrate. The solder on the pads are then melted during the reflow process to bond the two together to form an electrical connection.

Controlled Impedance Line: a signal line with constant impedance value.

Convection: heat is transferred due to hot air movement. It is often found in infrared (IR) radiation for reducing IR shadowing effect.

Correct-by-Design: a new design and layout process featured in today's MCM CAD packages by some of the advanced CAD vendors such as Cadence and Integraph to incorporate design and fabrication specifications into their software to make the designs a manufacturing success the first time.

Contraint-Driven-Design: in the area of MCM CAD packages, this is a new way of transforming electrical and fabrication constraints into physical - design guidelines to make component placement and routing more intelligent, yielding a design with better signal integrity and mechanical reliability.

Coplanarity: a differential value between the lowest and highest lead or pin of an IC or connector when it sits in a seating plane.

Coprocessor: an add-on to the processor to make it more powerful in computing.

Cp: process capability index. It is primarily used to indicate whether or not a process actually performs at the required capability.

Cratering: a pit is created under an ultrasonically bonded area of a chip that comes loose due to a large amount of electricity being applied to the wire bond.

Creep: the dimensional change of a material under load throughout the course of time.

Creepage: a movement of electricity across a dielectric material surface.

Crimp: to make an electrical connection between a conductor barrel and a cable by compressing the barrel.

Cross Coupling Noise: noise created as a result of cross capacitance and mutual inductance coupling between neighboring conductor traces in close proximity on the substrate.

Crosslinking: a process known as irreversible polymerization.

Crosstalk: as a result of capacitive or inductive coupling or both, signals from one conductor line leak into neighboring lines. It is usually found in two closely parallel conductor lines on the substrate.

Cryogenic: dealing with temperatures approaching absolute zero.

D

Defect Opportunity (DO): the possibility of causing a defect in a unit at a manufacturing location.

Die: a passive or active component that is unpackaged.

Dielectric Constant: the ratio between a charge stored with free space as the dielectric and the charge stored with the material in question as the dielectric.

Dielectric Strength: the ratio between the maximum voltage that an insulating material is capable of handling prior to the breakdown of voltage and the distance between electrodes. The unit is volts/mil.

Dissipation Factor: the ratio between R_c (the resistive component of a capacitor) and X_c (the capacitive reactance of a capacitor).

Dispersion: a process of distributing delicate particles in another substrate.

Distribution Layer: a special layer used for test and rework and route escape functions.

Dymalloy: a diamond/copper/silver composite material with excellent thermal property and best suited for heat sink and substrate use.

E

Elastomer: a material capable of stretching repeatedly to at least twice its length at room temperature and can return to its original length upon the release of force.

Electrical Resistance Test: a test used to measure the circuit resistances to ensure proper connections.

Embedded Controller: a product loaded with built-in intelligence so that it can have more computing power and user friendliness than an ordinary product.

Emissivity (Total Emissivity): for a thermal radiator such as metal under a given temperature, it is the ratio between the total radiation ouput of that radiator and a blackbody at the same temperature.

Emulsion: a light sensitive material used in the photofabrication of PCBs (printed circuit boards). Also coating the screens during thick-film printing.

Eutectic: a low temperature mixture with a temperature lower than all other mixtures with the same ingredients.

Eutectic Die Attach: the attachment of a die to the next level of interconnect (i.e. substrate) using an eutectic metal alloy preform.

F

Fanout of a Net: no. of pins connected together by a net -1.

Feature: an easy to recognize pattern on a part.

Feeder: a device for supplying components to the placement machines where components are oriented correctly and fed sequentially.

First Incidence Switching: it occurs when the first signal reaching the end of the line has enough voltage to switch the receiver.

Flash Memory: a type of memory that is nonvolatile, high density and requires no refreshing. Also, no battery is needed to power it.

Flexural Strength: the strength of a laminate material under bending.

Flip Chip: a semiconductor component such as die with solder pads or bumps as its terminations on one side is flipped over to make connections to the corresponding foot prints on the substrate.

Frit Glass: a type of glass that is finely ground and used in thick-film pastes. It will melt during firing and stick to the substrate as in the case of metal oxides.

G

Gerber Data: data containing X & Y coordinates, aperture lists, and operation commands are sent to a photoplotter to generate artwork.

Glass Transition Temperature: the temperature where a plastic transforms from a solid to a softened state. At this point, both electrical and physical properties of the plastic will degrade significantly.

Glob Top: a glob of encapsulation material such as silicone or epoxy is applied to the top surface of a bare chip after it has been wire bonded, inspected, and tested. This process will provide the bare chip with environmental protection.

Ground Plane: an embedded metal layer for connecting many points to one or more ground pins in the circuit.

H

HAST: a high temperature/high humidity bias test to accelerate the failure of parts.

Haze: the ratio of scattered power to incident power. It is primarily used in wafer inspection.

Heat Spreader: also known as heat sink. A metal object used exclusively for dissipating heat away from a hot object such as a semiconductor package that is in contact with it.

Hermetic Sealing: a process where materials are fused or soldered permanently together. It is typically applied to fusion between metals or metal and glass. Once sealing is done, no air or gases will be able to escape from or enter into the sealed area.

High Density Interconnect (HDI): a technology where a substrate is milled out to create a cavity for bare chips to go inside and then a Kapton film overlay is attached to the substrate and components to form the interconnections.

Hole-to-Wire-Ratio: the difference between the hole size and the wire or component lead size.

Host: a computer which has other terminals, smaller computers, and peripherals connected to it.

I

In-Circuit Test: the performance of electrical tests on individual components soldered onto their respective footprints on the board.

Inductor: a device capable of generating any voltages for the purpose of maintaining a constant current.

Inert Atmosphere: an atmosphere containing nitrogen or helium or a mixture of gases that will not oxidize or reduce when treating metals.

Insulation Resistance: a measure of resistance to current when a voltage potential is applied. The unit is megaohms.

Interconnection: point-to-point connection.

Interconnection Capacity: the total number of connections in a given level of packaging hierarchy.

Interstitial Via: a hole plated through with metal is used to connect two or more metal layers of a

PCB.

ISO: International Standards Organization. Due to its efforts, many quality assurance standards have been developed.

ISO9000: it covers the definitions of prominent quality management terms and is the recognized roadmap used for all quality standards.

ISO9001: it is more specific than ISO9000 because it establishes a quality system model for operations that encompasses the design, production, installation, and service of a product.

K

Kiln: a high temperature furnace used to fire ceramic substrates.

Knee Cut: used in resistor trimming where incremental changes in resistance value are medium.

L

L Cut: a type of resistor trimming that can only be applied to fairly wide resistors with low values where a tab is absent.

Lead Finish: a coating for protecting the leads from corrosion and ensuring that the leads are solderable.

Lead Resistance: DC resistance in the leads.

LPGA: laser programmable gate array. It has the

speed, density and low cost benefits of nested gate arrays and the fast sample and production turnaround cycle of FPGA.

Loss: a reduction in signal power during transmission from one point to another point.

Lossy Line: a transmission line encompassing skin effect, series resistance and dielectric conduction.

Low-Fired Ceramic Tape Process: a process pioneered by PMC (Pacific Microelectronics Corporation) where a low dielectric constant tape containing pre-drilled vias is laminated onto a standard alumina or beryllia substrate. Then, the substrate is fired in a high temperature infrared furnace between 850 and 900°C.

M

MCM-C: an MCM with either a low K or co-fired ceramic substrate.

MCM-D: an MCM with conductor patterns deposited onto a silicon or metal or ceramic substrate using thin film deposition similar to IC technology.

MCM-L: an MCM with a high density laminated printed circuit board.

MCM-L/O: MCM laminate/overlay. A high density packaging technology in which all power, ground and off-module connections are in the PCB so that only chip-to-chip connections are in the overlay to enable shielding the interchip signals from any potential high current paths.

MCM-Si: an MCM with conductor patterns deposited onto a silicon substrate using thin film deposition similar to IC technology.

MCP: multichip packages or few-chip packages. A technical term used only by National Semiconductor Corporation in which five or less chips are connected to the substrate using either wire bonding or TAB method. The entire construction is then protected by either a molded encapsulant or an affordable ceramic package.

Mean Time Between Failure (MTBF): the average time between a unit that has failed and the next unit that will fail, in hours.

Microstrip: a construction that is often used in high speed digital design where a rectangular conductor sits over a reference plane, separated by a dielectric material. The reference plane is either a power or ground plane.

Mini Ball Grid Array (mBGA): a technology developed by Sandia National Lab in which the overall package has a size very close to the die.

Multichip Module (MCM): a newer form of packaging technology where several bare chips are assembled onto the substrate using either wire bonding, TAB or flip chip method. Other surface mount components such as capacitors and resistors can be soldered onto the substrate as well. Once component assembly is completed, electrical connections are made between all components and the substrate. The overall package can be either globbed or resin molded depending on the final field applications.

Multilevel Circuit: a circuit containing layers of conductor separated by dielectric layers between each conductor pair.

Multiterminal Net: a net with multiple nodes.

Multiwire Technology: an extremely high density interconnect technology combining design rule-based software and a fully automated discrete wiring manufacturing process to yield a thinner package with predictable transmission line characteristics.

Mutual Capacitance: the capacitance between two conductors where all other conductors are grounded electrically.

N

Net: a set of two or more pins connected electrically.

Newbie: a new user on the Internet.

O

Open Architecture: a system that can be expanded by adding more elements to it.

P

Pareto Principle: the 80/20 rule in which 80% of the work gets handled in the last 20% of the time.

Peak Performance of a System: it can be obtained if the relative performance of a system reaches 100%, meaning that the number of minimum cycles and the actual number of cycles are equal. Another way of looking at it is that memories can be accessed with the minimum number of cycles with zero wait states.

Percentage of Reflected Signal: the ratio between the difference between the termination and trace impedance and the sum of the two.

Permanganate Etch: a process where the epoxy surface is first roughened to enhance the adhesion of an electroless copper, then a desirable copper thickness is achieved via electrolytic plating and followed by subtractive etching to obtain the correct circuit patterns.

Photomask: a patterned screen is used to shield and expose certain areas of a photosensitive substrate surface to light so that the substrate surface can become a mask.

Pick-and-Place: an assembly process where components such as ICs and other devices are oriented and placed properly onto their corresponding pads on the substrate before soldering.

Plating: there are two types of plating. One is called electroplating and the other one is known as electroless plating. In electroplating, the substrate is formed with conductive films such as gold or copper. In electroless plating, there is no need for electrical conduction during the plating process, therefore, insulating substrates can be used. Metals of choice are gold, copper and nickel.

Plug-and-Play: the latest technology in PCs where users can simply insert a card or peripheral into a desktop system and get the system running without going through difficult hardware and software modifications.

Plunge Cut: also known as serpentine cut. It is frequently used in resistor trimming where large resistance changes are expected.

Preconditioning: a process where packages are subject to ambient humid environments before solder reflow. Packages are surrounded by moisture for varying times and once a particular moisture level is detected inside them, known temperature ranges and profiles can be used on the packages to determine their crack resistance levels.

Preheating: heating a compound before molding or casting so that it is more reliable.

Prepreg: a fiber filled with B stage resin.

Propagation Delay: the time delay between the input and output of a signal. The unit is nanoseconds per foot of conductor length.

Q

Quality Test Action Group (QTAG): it was formed to promote IDDQ testing, define a standard, and investigate IDDQ measurement techniques.

R

Random Access Memory (RAM): information can be retrieved or written into a memory randomly by choosing its address. A specific delay in time will be involved in the read/write operation.

RAS: it is used to drive the RAS inputs of either a VRAM (video RAM) or a DRAM (dynamic RAM). It is also known as registrar accreditation board.

Reactive Ion Etching (RIE): reactive ion etching. A plasma etching process where materials are removed by chemical reaction and active ion bombardment under a high electric and low pressure field.

Reduced Instruction Set Computing (RISC): it was developed to replace CISC to be the computing technology for MPU design. With its reduced instruction sets and sophisticated compiler algorithm, the speed of processing can be improved significantly.

Reflection Coefficient: the ratio of the difference between the load impedance and characteristic impedance of a signal line to the sum of the two.

Registration Corner: a corner created by the intersection of two sides of a substrate. It is primarily used as the reference point for all measurements.

Regular Pad: a positive pad with a regular shape, e.g., a circle.

Relative Performance Ratio: this applies strictly to system performance. It is the ratio between the number of minimum cycles and the actual number of cycles.

Ring Frame: it comes in one piece and is normally made from Alloy 42 or extruded Kovar tubing. It comes in all different shapes and once attached, it makes up the sidewalls of a package. It must be plated electrolessly with nickel to make it easier for brazing.

S

SBGAÔ: super ball grid array. A high power, thermally and electrically enhanced BGA package in which the bare chip is directly mounted on an exposed copper heatsink.

Scaleable: to scale up or increase the performance of silicon technology.

Scan-sequential: the combination of full-scan algorithm and reduced overhead in sequential logic.

SFROMª: smart flash ROM by EUROM Flashware Solutions Ltd., an Israeli company. It is the ideal device for non-volatile memory because of its re-programmability, large flash storage capacity, continuous write and rewrite functions and the absence of UV erase.

Shorting Bar: used in resistor trimming where the incremental changes in resistance value are large.

Signal-to-Noise Ratio: the ratio between the root mean square signal power and the root mean square noise power at a radar receiver output.

Silk Screening: a printing process where a paste is squeezed through the open areas of a screen and is deposited onto the substrate to form conductor patterns.

Silver Migration: the movement of ions in silver from one area to another area.

Single Layer Circuit: a circuit containing only one conductor layer. Typically, most of the simple resistor networks can be classified as single layer circuits.

Singulation: a process used for dicing a wafer into individual die.

Sintering: a process where a metal powder is heated under pressure and all its particles bond together to form a mass.

Skin Depth: the distance at which the current density equals 0.368 of its value at the conductor surface.

Skin Effect: an effect where a conductor resistance is increased under microwave frequencies due to the concentration of current at the conductor surface.

SLICC: a small form BGA packaging technology developed by Motorola's Land Mobile Products Sector group.

Slump: a spreading phenomenon found in the screen-printed thick film composition prior to drying.

Slurry: a mixture where suspended solids are found in liquid.

Small Outline (SO): a plastic molded package having gull wing leads.

Smart IC: an integrated circuit component possessing the ability to perform extra logic functions.

Solderability: the wetting and bonding ability of a metal when in contact with solder.

Solder Bridge: as a result of solder connecting two conductors, an electrical short occurs between these conductors.

Solder Bumps: solder balls used for face-down bonding to the corresponding pads on the next level of interconnect (i.e. substrate).

Solder Pot: a pot filled with solder.

SPARCÔ: scaleable processor architecture. An industry standard 32-bit RISC microprocessor architecture developed by Sun Microsystems.

Sprue: a conduit for molten plastic or metal to flow into the mold.

State-of-the-Art: the latest and greatest technology.

Stripline: a construction that is often used in high speed digital design where a rectangular conductor sits between two reference planes, separated by dielectric layers.

Stub: short portion of signal line.

Substrate Efficiency: the percentage of the substrate occupied by silicon.

Surface Finish: the nonuniformities observed in the surface of a solid material. The unit is micro-inch per inch.

Synchronous DRAM (SDRAM): a higher performance part than the JEDEC standard DRAM. It operates its internal control interface from a single synchronous clock and has burst-mode and on-chip interleave functions.

T

Tape Automated Bonding (TAB): a technology where a chip's bonding pads are bonded with their corresponding etched leads using thermocompression. The leads are supported on a flexible tape.

Tape & Reel: a reel consisting of components that are transported via a tape with cavities.

Thermal Electrical Cooler (TEC): a solid-state device that pumps heat using the Peltier effect when external power is applied to it.

Thermal Conductivity: the transmission rate of heat through a material body.

Thermal Relief: a negative pad on a positive copper area or a positive pad on an embedded metal layer used for power or ground distribution.

Thermount: an aramid-based nonwoven material by Du Pont. It has a smooth mounting surface that is light and thin, and a low dielectric constant (4.1 @ 1MHz). Precise drilling of small holes can be achieved to yield certain thermal coefficients of expansion.

Thermosetting: a process where a plastic is hardened by heat into a permanent shape. Once a plastic is thermally set, it cannot be re-melted but can be softened by heat.

Thermosonic Wire Bonding: a process where a gold wire is bonded to a die or a substrate pad when ultrasonic power is applied to a capillary of a thermocompression bonder. The bonding temperature is around 150°C.

Thermal Shock: an accelerating temperature test where components or equipment are subject to an alternating temperature fluctuation between low and high temperature. This process will cause early part failures.

Thick Film Circuit: a circuit containing layers of conductor patterns which are screen printed and fired onto a ceramic substrate in consecutive orders.

Thick Film Resistor Chip: it is obtained by slicing a thick film resistor substrate containing multiple images.

Thick Film Resistor Array: either a substrate or chip containing only conductors and resistors in a single layer construction.

Thick Film Resistor Substrate: a substrate containing screen-printed thick film resistors as its circuit.

3D-MCM: a construction where several MCM modules

are stacked vertically.

Time-of-Flight: the speed of light in a material. It is the ratio of the product between the length of the signal trace and the square root of the dielectric material's constant to the speed of light (3 x 10^8 m/s)

Time-of-Flight Delay: the propagation delay of a signal when it travels from the source to the sink terminal on a net.

Top Hat: used frequently in active trimming where large resistance changes are expected.

Transient: an unexpected change in the system due to large current surge on either the power or signal line, resulting in possible component and equipment failure.

Trimming Tabs: used in trimming resistors with higher resistance values. With this method, the operator could zero in on the target values more precisely.

V

Vent: a miniaturized hole in a mold for air to escape as plastic molding material is being poured into the mold, resulting in no bubbles or air holes in the cured plastic part.

Via Starvation: a router having difficulty of finding a via when it needs one, making the completion of a particular connection extremely difficult.

Vision: used with assembly machines to help place components accurately on the substrate.

W

Waffle Tray: also known as matrix tray. A tray filled with rectangular cavities used for housing and protecting IC devices.

Warpage: the dimensional distortion of a substrate from its flat surface.

Wave Soldering: a process where a wave of moving solder is brought in contact with multiple solder joints for a short time. All the solder joints are then soldered at the same time.

Wetting: during soldering, the molten solder flows over a metal surface uniformly as flux and heat are applied to it.

Wicking: a desoldering process where a prefluxed braid of stranded wire is used to remove solder from a solder joint by placing it above the joint and heating the joint with a soldering iron.

Wire Bonding: an assembly method used to mechanically connect a substrate pad and a semiconductor pad or leadframe together via a fine metal wire, usually gold or aluminum. There are three wire bonding methods - thermosonic, ultrasonic and thermocompression bonding.

Wire Sag: the inability of a wire to maintain its loop as defined by the wire bonding path between two pads.

X

X-Ray Diffractometer: equipment used primarily in the measurement and recording of X-ray diffraction patterns and crystals to analyze their structures.

Y

Yield: the least amount of stress applied before the material begins to deform without any additional increase in load.

Young's Modulus: the ratio between nominal stress and its corresponding compressed stress that is below the physical limit of the material.

Z

Z-Axis: in a three-dimensional zone, it is the axis that is perpendicular to both the x and y axis.

Zener Diode: a semiconductor diode with two layers that functions in the reverse-bias breakdown condition.

Zero Wait State: no clock cycles are added to the minimum number of cycles to accommodate certain memory devices such as DRAMs for their slow memory access requirements.